ANCIENT EGYPT: SHIPPING AND TRADING LESSONS FROM HISTORY

By: Mustafa Nejem

CONTENTS

INTRODUCTION

When I was unaware of the depth of ancient Egypt's history, it often appeared to me as a land ruled by a bunch of arrogant, wealthy kings, where the primary occupation was mummy-making and the practice of mysterious witchcraft, destined to reawaken centuries later. This is the common perception of ancient Egypt among ordinary folks, largely influenced by Hollywood movies.

As I began my studies, I soon realised that while the incredible practice of mummification was a fascinating aspect of ancient Egypt's civilisation, it was just one facet of a much broader and sophisticated culture. The people of this land were civil, highly educated, and wise. From 3100 BCE to 332 BCE, this civilisation thrived on intellect and ingenuity rather than relying on modern technology. Trade, too, played a pivotal role, especially with neighbouring civilisations, enabling them to acquire precious items such as metals and cedarwood. Like in other aspects of their lives, they excelled in trade, proving that their entrepreneurial skills and spirit should never be underestimated. In the pages of this book, we will delve deeper into their trading prowess and explore the valuable lessons modern entrepreneurs can glean from their remarkable achievements.

Chapter 1

LEVERAGING
GEOGRAPHICAL ADVANTAGES

Few civilisations in human history have been as fascinating and resilient as ancient Egypt. For ages, scholars and aficionados have been fascinated by its imposing pyramids, stately pharaohs, and complicated hieroglyphics. Beyond the mysterious tombs and the golden sands of the desert, however, is a vibrant ancient economy intricately connected to the busy ports along the Mediterranean coast and the Nile's waterways. We're glad you're here as we reveal the fascinating story of ancient Egypt's river and coastal trade. In this chapter, we will examine how the Nile River, which provided life, and the key coastal ports, emphasising the renowned Port of Alexandria, acted as the economic arteries that supported this magnificent civilisation as we set out on this voyage.

River and Coastal Trade

As per studies, it is found that the establishment of land and sea trade routes gradually connected the ancient Egyptian civilisation with other civilisations in ancient India, the fertile region of Arabia, and Sub-Saharan Africa. This was the basis of ancient Egyptian trade as researchers found that the trade via river and coastal regions was too common for them. Egyptians utilised the Nile River and Mediterranean coastline to do trade.

Utilizing the Nile River and Mediterranean coastline

Egyptians were versatile traders who usually used ships that crossed the Nile and the Mediterranean Sea, which the Egyptians used for commercial purposes. They also participated in trade that took place over land. Oases and locations along the Nile and other important trade routes were used to establish way stations. Although there was no money at the time, the government was responsible for collecting things in a centralised location and distributing them.

a) The Nile River: A Natural Trade Highway

When the Greek historian Herodotus claimed that the ancient Egyptians' land was "given to these people by the river," he was alluding to the Nile, whose waters were important to developing one of the world's earliest great civilisations. Herodotus's writing is considered one of the earliest examples of historical writing. Ancient Egypt benefited from the Nile by receiving abundant water and soil for irrigation, as well as a way of transporting materials for construction projects. The Nile runs northward for 4,160 miles from east and central Africa to the Mediterranean. Because of its life-giving waterways, cities could emerge in the middle of a desert.

For those who lived along the banks of the Nile to reap the benefits of the river, they needed to develop strategies for surviving the annual floods that the Nile caused. They also invented new techniques and technologies in various fields, ranging from agriculture to the construction of boats and ships. Even the pyramids, those enormous architectural wonders that are probably the most iconic relics of Egyptian civilisation, were built with the help of the Nile.

The immense river had a significant impact on the day-to-day lives of ancient Egyptians and how they viewed themselves and the world around them. Their religion and culture developed according to statements made by Lisa Saladino Haney, assistant curator of Egypt at the Carnegie Institution of Natural History in Pittsburgh, who quoted on the institution's website that the Nile was "a critical pulse that brought life to the desert." In his book published in 2012 titled "The Nile," an Egyptologist states that "without the Nile, there would certainly be no Egypt."

Nile: A source of rich Farmland:

The present name for the river Nile originates from the Greek word "Nelios," which means "river valley." However, the ancient Egyptians referred to it as Ar or Aur, both of which mean "black." This was a reference to the rich. Dark silt that the water of the Nile carried from the Continent of Africa northward and dropped in Egypt as the river broke its banks every year in the late summer. Because of this influx of water and nutrients, the Nile Valley was transformed into fertile farmland, allowing Egyptian civilisation to flourish despite its location in the desert.

According to Barry J. Kemp, author of Ancient Egypt: Anatomy of a Civilization, the heavy layer of silt that settled in the Nile Valley "changed" what might have been a scientific curiosity, a version of the Grand Canyon, into a heavily settled agricultural country." The ancient Egyptians placed such significance on the Nile that the first phase of the flooding season was chosen to mark the beginning of the year on their calendar. In Egyptian religion, a god named Hapy was revered. Happy was portrayed as a rotund man with either green or blue skin and considered the God of flooding and fertility.

According to the Food and Agriculture Organization of the United Nations (FAO), ancient Egyptian landowners were one of the first cultures to undertake agriculture on a big scale. Ancient Egyptian farmers grew food crops, including wheat and barley, and industrial crops, such as flax, used to make garments.

Ancient Egyptian farmers developed a method known as basin irrigation to make the most efficient use of the water provided by the Nile. They excavated channels to route the rivers of flood water into the basins, where it would sit for a month until the soil was soaked and suitable for sowing. They created networks of earthen banks to make basins. Arthur Goldschmidt states, "It is obviously tough if the land on which you have built your home and grow your food gets flooded by a river every August and September," as the Nile used to do before the Aswan High Dam. Goldschmidt is also the author of "A Brief History of Egypt." The ancient Egyptians had to use their inventiveness and likely went through a lot of trial-and-error experimentation to shift and store some of the Nile's waters. They did this by building dikes, canals, and basins. Ancient Egyptians constructed nilometers, which were stone columns with markings that indicated the water level. These nilometers allowed the ancient Egyptians to determine whether they would be subject to perilous floods or low waters, which could lead to a bad crop.

Nile: A vital transportation route

The ancient Egyptians used the Nile River as a crucial transportation route, in addition to its role in agriculture cultivation. They were able to become adept boat and shipbuilders as a result, and they developed both larger wooden vessels with sails and oars capable of sailing greater distances and smaller skiffs made of papyrus reeds linked to wooden frames. Old Kingdom artwork from 2686 to 2181 B.C. showcases boats carrying animals, vegetables, fish, bread and wood.

This period of Egyptian history spans from 2686 to 2181 B.C. The Egyptians placed such a high value on boats that they even placed them in the tombs of deceased kings and officials. These boats were built occasionally with such precision that they were seaworthy and might have been used for navigation on the Nile.Mediterranean Coastline: Gateway to International Trade

History of the Mediterranean Sea:

The heritage of the Mediterranean Sea is a record of people and cultures communicating with one another from the surrounding areas. Some of the earliest human civilisations to be structured around it were the Egyptian, Phoenician, Greek, and Roman civilisations. They had a significant impact on the past and the very life of the cultures across the Mediterranean. As a result, they are one of the keys to understanding the evolution of Western civilisation as it is currently known to everyone. The relationships that the residents of the Mediterranean basin had with its terrain were not, however, at all times on an equal footing.

It has been thousands of years since the firm that was most ingrained in a region (in which spatial awareness developed in the fantasy) was able to manage enormous empires. In that time, the world has changed significantly.

In the case of Egypt, the road of the Mediterranean was never absent; nonetheless, it wasn't until much later that it was finally conquered. The civilisation, which owed its continuation, persistence, and peacefulness to its unifying elements – Pharaoh, Maat, religion, language– and the peace that guaranteed its borders, gradually lost a significant amount of its originality:

tradition, which had been so well preserved for centuries, mixed with the traditions and cultural traits of other civilisations.

In this way, this "cultural exchange" was mutual, as Egypt also left to the rest of the globe a set of messages and practices that are particularly significant. The opening to the world permitted access to knowledge of new political, religious, and mental features (for instance, the language field has been enriched with new words). However, the opening to the world also allowed access to knowledge of new terms. A significant contribution that Egypt made to Western civilisation was the establishment of the word as a model for the creative process.

Ptah, the deity of creation in Memphis, conceived of the world in all its myriad forms within his heart and brought them into existence through the creative power of the operative word. Although it is typically associated with the biblical text and is located in a specific historical, geographic, and temporal context, the doctrine of the creative verb originates in a very different time and space: the Nile Valley. In this region, the various priestly castes offered a variety of explanations for the widest variety of occurrences. The legacy of Egyptian civilisation endured throughout the course of history, making its way across the Mediterranean to numerous meeting places such as Syria, Phoenicia, Palestine, and Anatolia before being intertwined with our Western tradition via two very different channels: the authors of classical antiquity and the Bible.

Location:

The Mediterranean Sea is a body of water found in the middle of Europe on the one side and Africa on the other. The sea is linked to Western Asia to the east and the Atlantic Ocean via the "Strait of Gibraltar" to the west. Western Asia is connected to the Mediterranean Sea to the east. Throughout the extensive history of Ancient Egypt, many civilisations have flourished and perished in and around the Mediterranean.

The Nile River eventually discharged into the Mediterranean Sea. Even in the early Predynastic period of Egypt, the river had already become an important commercial thoroughfare. Even though Egypt was already trading with other Mediterranean nations by the 4th millennium BCE, the country did not actively participate in maritime commerce in the Mediterranean Sea until much later. Ancient Egypt cultivated its distinctive culture over many decades thanks to the sea's role as a barrier that kept it isolated from the outside world.

In Egypt during the Protodynastic period, An Egyptian community had been created, and it had its roots before the First Dynasty. This colony was located in southern Canaan. Canaanite regions like Arad, En Besor, Rafiah, and Tel Erani were used to create Egyptian pottery to export back to Egypt. Narmer had his name stamped on the vessels that were produced. In 1994, archaeologists discovered an inscribed ceramic fragment with the serekh symbol of Narmer dating back to approximately 3000 BCE. The shard was determined to be a piece of a wine jar transported from the Nile Valley to the Mediterranean through mineralogical research.

Mediterranean Sea as Trading Port:

The ancient Egyptians had a working knowledge of shipbuilding at 3000 BCE or possibly earlier. They used woven straps to lash the wooden boards together, and reeds or grass packed in between the planks served to seal the seams. The ancient Egyptians knew how to put the wood planks into a ship's hull and utilised woven straps.

According to the Archaeological Institute of America (AIA), the longest ship ever dated, which measured 75 feet in length and dated back to 3000 BCE, may have belonged to Pharaoh Aha. The Palermo stone references King Sneferu of the 4th Dynasty of Egypt sending ships to Lebanon to acquire high-quality cedar. Egyptians bring back massive cedar trees in one of the scenes depicted in the pyramid of Sahure, which dates back to the Fifth Dynasty. The name Sahure was found imprinted on a small piece of gold attached to a chair from Lebanon, while the 5th dynasty cartouches were discovered on stone objects from Lebanon.

In several other scenes in his temple, bears native to Syria are depicted. The Palermo stone refers to journeys to the Sinai Peninsula and the diorite quarries north of Abu Simbel.

An ancient "Suez" Canal is reported to have been begun by the mythical figure Sesostris, who was most likely either Pharaoh Senusret II or Pharaoh Senusret III of the Twelfth dynasty of Egypt. This canal was supposed to connect the River Nile to the Red Sea. Aristotle, Pliny the Elder, and Strabo are three ancient authors supporting this claim. "One of their kings attempted to build a canal to it (for it would have been to their great advantage for the entire region to have become connected to it)," it is said that Sesostris was the first of the ancient monarchs to try to navigate it. However, he discovered that the sea was higher than the land. Therefore, he stopped making the canal first, and Darius did so after him to prevent the water from the river and the sea from becoming contaminated.

Furthermore, following that is the Tyro tribe, as well as, on the Red Sea, the harbour of the Daneoi, from where Sesostris, the ruler of Egypt, intended to construct a ship canal to where the Nile runs into what is known as the Delta; this is a distance of more than 60 miles. The distance between these two points is over 100 kilometres. In later times, King Darius of Persia had the same idea, and still another ruler named Ptolemy II who constructed a trench that was around 35 miles long, 100 feet broad, and 30 feet deep—all the way to the Bitter Lakes. In the end, the canal was constructed. Darius I of Persia, the ruler responsible for the conquest of Ancient Egypt, constructed a canal connecting the Mediterranean Sea and the Red Sea. The journey across Darius' canal took four days, even though it was wide enough to allow two triremes to pass between themselves with their oars extended.

Real-Life Example: Port of Alexandria

According to the experts, Egypt's recent plan to rebuild its largest port, the Port of Alexandria, is a significant step toward becoming a major hub for international trade and transportation. Egypt recently announced this plan. By constructing new docks, roadways, and other infrastructures, the Egyptian government is working toward integrating several neighbouring ports along the Mediterranean Sea into a single comprehensive and highly modern port.

Ahmed al-Wakeel, head of the General Union for Commercial Chambers in Egypt, said, "The new project is part of the country's strategic plans for promoting the transportation sector by establishing developmental corridors and logistics zones and linking numerous ports with railways and roads." These are all goals that the country has set for itself to help advance the transportation sector.

He told Xinhua that the giant Alexandria port will be a big leap for the country's transportation and logistics, and he added that the highly built substructure will enable the port to serve international trade and economy, which will help Egypt attract more investment. He said the highly developed substructure will also help Egypt attract more tourists. According to the

Minister of Transport Kamel El-Wazir, after the project was completed in 2021, the port of Alexandria will become one of the major ports in the Mediterranean region, with larger and deeper docks. The minister went on to say that the port had already constructed a garage that was five stories tall, 15,000 square meters in size, and could hold 4,000 vehicles. In addition to this, it is working on the construction of a station for the trade of wood that will be able to accommodate ships weighing up to 70,000 tons. Along the 24.9 kilometres of coastline, the integrated port will consist of 87 platforms, and its construction is anticipated to cost approximately 12 billion Egyptian pounds, equivalent to approximately 764 million USD.

Al-Wakeel mentioned that Egypt has a great readiness to make full use of its geographic advantage to develop the logistics and trade industries. He added that the project, once finished, would boost the competitiveness of the Alexandria port in the area. Al-Wakeel also mentioned that Egypt is strongly willing to promote logistics and trading industries. According to Kareem al-Omda, a professor of the economy at the Arab Academy for Science, Technology and Maritime Transport, the project will also improve the country's transit and re-exportation services. "The new projects will greatly increase trade activities, which will then raise the state's revenues from the sale of foreign currencies as well as supplies of tariffs and taxes," al-Omda explained further.

According to Zynab Nawar, a professor of the economy at the British University in Cairo, the new project will greatly enhance the port's capacity while reducing the time that ships are delayed. "The process of further deepening the ports, adding many platforms, and building ground roads and railways as well as river means of transport will shorten the time it takes to transfer commodities and reduce costs, resulting in decreased product prices," the economic expert stated. "Additionally, further deepening the ports will shorten the time it takes to transfer commodities and increase the number of platforms. Nawar noted that enhancing the quality of Egypt's transportation infrastructure in general will lower the cost of transportation and, eventually, the cost of manufacturing, making Egyptian goods more competitive globally. According to government data, the port, constructed in 1835, is the most important in Egypt and is responsible for sixty per cent of Egypt's imports and exports.

Anecdote: The Story of Imhotep's Trade Expedition
The ancient Egyptian builder, mathematician, physician, astrologer, poet, and priest, Imhotep, was also Chief Minister to the Pharaoh Djoser. He was a genius in several other fields as well. The name Imhotep means "the One Who Comes in Peace." He was an ancient Egyptian king. Imhotep was a commoner when he was born, but he rose through the ranks to become the vizier of King Djoser and was in control of erecting the tomb of King Djoser at Saqqara. Imhotep, an Egyptian trader, capitalised on the Nile's water route to transport grain from Upper Egypt to the bustling market in Memphis. This not only contributed to regional food security but also boosted Imhotep's trade profits.

Anecdote: The Rise of Alexandria
Alexander the Great established the city in 332 BCE, shortly after his campaign in Persia began. He intended for it to serve as the capital of his newly established Egyptian kingdom and a naval base from which he could exercise authority over the Mediterranean. The amount of water given by Lake Mary, which was at the time fed by a spur of the Canopic Nile, and the good anchorage afforded offshore by the island of Pharos were two of the deciding factors that led to the selection of the location that includes the old city of Rhakotis (which dates back to 1500 BCE). Rhakotis was established at this location. Cleomenes, Alexander's viceroy in

Egypt, continued the construction of Alexandria after Alexander had departed the country. After Alexander died in 323 BCE, the empire began to fall apart, and Ptolemy I Soter, Alexander's viceroy, assumed control of the city.

Ptolemy I Soter also established the dynasty that would later bear his name. In the year 332 BCE, shortly after the outbreak of Alexander the Great's war in Persia. Alexander founded this city. He planned for it to be the capital of his newly founded Egyptian kingdom and a naval base from which he could exert dominance over the Mediterranean Sea. He also intended for it to serve both of these functions simultaneously.

Moreover, two of the deciding factors that led to the choosing of the location that includes the old city of Rhakotis (which dates back to 1500 BCE) were the quantity of water provided by Lake Mary, which at the time was fed by a spur of the Canopic Nile, and the good attachment afforded offshore by the island of Pharos. This place became known as Rhakotis after it was founded. After Alexander had left Egypt, the construction of Alexandria was carried on by Cleomenes, Alexander's viceroy there. Ptolemy I Soter, Alexander's viceroy, took leadership of the city after Alexander died in 323 BCE, which marked the beginning of the empire's decline and subsequent breakup. In addition, Ptolemy I Soter laid the foundation for the dynasty that later came to be referred to by his name.

Anecdote: Building the Koptos-Nile Canal

The Koptos Nile Canal, also known as the Koptos Canal or the Coptos Canal, was an important artificial waterway in ancient Egypt that played a major part in the commercial and transportation network of the country. Other names for this canal include the Coptos Canal and the Koptos Canal. Through the Koptos Nile Canal, the Nile River was ultimately able to reach the Red Sea.

It began close to the ancient city of Koptos, now known as the contemporary city of Qift or Qeft and is located in Upper Egypt. In ancient times, Koptos was an important trading town, extending eastward to the Red Sea coast. The Koptos Nile Canal was constructed primarily to ease the flow of commerce between the Nile Valley and the Red Sea.

Furthermore, it played an important role in the transportation of commodities to and from the ports located on the Red Sea, most notably the port of Berenice (which is now known as Marsa Berenice), which functioned as a port of entry for the Indian Ocean and was a hub for trade with the east. The construction of the Koptos Nile Canal was an event that had enormous historical significance. It allowed important items to be transported between Egypt and the faraway nations of the Arabian Peninsula, the Horn of Africa, and the Indian subcontinent.

These goods included precious stones, metals, spices, and other goods. Notably, the well-known expedition that Queen Hatshepsut led to the Land of Punt in the 15th century BCE is thought to have used the Koptos Nile Canal to bring the unique items and resources they acquired back to Egypt. Both the building of the canal and its ongoing maintenance were significant undertakings.

Engineers and labourers in ancient Egypt had to overcome obstacles such as fluctuating water levels, seasonal flooding, and the need to maintain a consistent flow of water from the Nile to the canal to complete the project. The canal needed to be maintained and dredged regularly. The canal was allowed to deteriorate over time until it was finally closed off and forgotten about. It is possible that changes in trade routes, upheavals in political power, and the erection of newer canals, such as the Canal of the Pharaohs (a forerunner to the contemporary Suez Canal), contributed to the canal's demise. These waterways fulfilled purposes comparable to those of the modern Suez Canal. Even though the Koptos Nile Canal is not in operation now, its importance in history cannot be understated because it was an early engineering marvel and a commercial route during the ancient world. Therefore, it contributed to Egypt's long history as a centre of trade and commerce and served as a foundation for later canal-building attempts in Egypt. Additionally, it created the groundwork for future canal-building endeavours in Egypt.

The legacy of ancient Egypt's coastal trade

The legacy of ancient Egypt's river and coastal trade shows how geography, creativity, and human ingenuity shaped human history. After exploring the life-giving Nile and the bustling Port of Alexandria, we are reminded that a civilisation that mastered the trade and navigated antiquity's waters with grace lies beneath the sands of time. Nile River, ever-sustaining, was the birthplace of Egyptian agriculture and the centre of trade. Its yearly floods nourished the soil, enabling abundant grain and papyrus harvests for wealth and wisdom.

Its canals and boats connected Upper and Lower Egypt, promoting trade and unity. At the crossroads of continents, the bustling Port of Alexandria was also on our itinerary. The Pharos Lighthouse guided mariners and a diverse civilisation brought their valuables here. Alexandria became a bustling market and intellectual centre when ships from Rome, Greece, Persia, and India brought goods, ideas, and art. Beyond the wealth and bustling quays, ancient Egypt's river and coastal trade gives us lasting truths. It emphasises how natural resources and geography shape civilisations.

The intricate irrigation systems controlling the Nile and Alexandria's harbour engineering wonders demonstrate ingenuity. With its massive lighthouse and famed library, the Port of Alexandria proves that trade is about exchanging cultures and ideas as well as things. This shows how a city may become a beacon of enlightenment and a repository of human knowledge beyond its commercial role. Our departure from the Nile and Alexandria's lively quays leaves us with the profound legacy of ancient Egypt's river and coastal trade, which means resilience, creativity, and human enterprise.

Conclusion

As concluded, ancient Egypt had a fascinating economy tied to the Nile River and the Mediterranean coast, as well as geographical benefits. They traded with places like India and Arabia through land and sea routes. The Nile was vital for farming and transporting materials. Not only this, they even used their sharp wits to invent clever methods like basin irrigation and nilometers to control the river's power. The Mediterranean coast, especially Alexandria, was a hub for international trade, connecting Egypt with other ancient civilisations like Greece and Rome. Therefore, the exchange influenced Western civilisation. Hence, ancient Egypt's trade legacy shows how they used their resources and creativity to thrive.

Chapter 2

Diverse Fleet Management

One of the world's most famous Egyptian civilisations thrived due to its transportation infrastructure. This ancient land relied on the Nile River for food, trade, culture, and exploration. This extraordinary transportation network relied on a diversified fleet of vessels designed for specialised functions. This journey explores ancient Egyptian fleet management and the Felucca and Khufu ships. We'll learn how this diversified fleet drove the civilisation's economic and cultural growth, demonstrating how fleet management may change an ancient culture.

Type of Vessels Used:
Egyptians didn't build highways throughout their empire. It wasn't necessary. The Nile River was nature's motorway through their kingdom. Most Ancient Egyptian cities were along the Nile. The Egyptians used the Nile for shipping and transportation from the start. They mastered boatbuilding and river navigation. They used boats, cargo ships, funeral boats, and wooden boats.

1. Small boats for local river trade:
Ancient Egyptians used early boats and wooden boats for local travel.

Early Boats:
Ancient Egyptians made miniature papyrus boats. They were easy to build and good for fishing and short outings. Oars and poles guided most papyrus boats, which were small. Traditional boats were long and slender with protruding ends.

Wooden Boats:

Egyptians eventually made wooden boats. Egyptian acacia and Lebanese cedar were used. They also started using a huge sail in the middle of the boat to catch the wind upstream. Ancient Egyptians fashioned wooden boats nail-free. Many short boards were joined together and connected with ropes to make boats. A big rudder oar at the back of the ship guided it.

Funeral vessels:

The Egyptians thought a boat was required to travel to the heavens in the afterlife. A tiny boat model was occasionally interred alongside a person. The tombs of rich Egyptians and pharaohs frequently had full-size boats. In some form, the tomb of Pharaoh Tutankhamun included 35 boats.

The Felucca, a Versatile Boat

The journey to Egypt offers felucca rides on the Nile like the Pharaohs did—long shadows from the mudbrick farmhouse cover pale, sun-scorched sand. Dark date palm shadows appear on the horizon. A late afternoon gust breaks the river surface, throwing water high onto the muddy bank. The ancient wooden boat travels quickly from the land's edge, its off-white sails full and stretched with momentum. Boatman covers his brow. He stares into the distorted, sun-reflected space between sea and sky. His face has all the proper creases and wears like a favourite leather shoe. Our modest boat bounces and rolls like a hungry dog's tail at dinnertime as he smiles and waves at every passing boat.

Large ships for long-distance maritime trade

Ancient Egyptians used cargo ships to travel internationally.

Cargo Ships:

The Egyptians mastered the art of creating substantial, durable cargo ships. To conduct business with other nations, they sailed these up and down the Nile and into the Mediterranean Sea. There could be a lot of cargo on these ships. Huge stones weighing up to 500 tons were transported by ships from the rock quarry to the construction site of the pyramids.

Military Boats:

Steve Vinson of Indiana University wrote: "Boats and warfare date back to the Predynastic Period. The ivory Gebel el-Arak knife handle, possibly from Naqada II/Gerzean times, shows two rows of boats of opposing designs underneath two registers of men fighting. The Gebel el-Arak knife handle was once thought to support the theory that a "Dynastic Race," possibly from Sumer, infiltrated Egypt around 3100 B.C. because the boats in the upper row strongly resemble Mesopotamian craft depicted on contemporaneous representations.

A Magnificent Example: The Khufu Ship:

An ancient Egyptian ship called the Khufu Ship was found in 1954 by a group of archaeologists under the direction of Kamal el-Mallakh. It is thought that this ship, discovered buried in the sand close to the Great Pyramid of Giza, was constructed during the rule of Pharaoh Khufu (2589–2566 BC). It has been called a "masterpiece of woodcraft" and is one of antiquity's oldest and best-preserved vessels. The ship comprises 1,224 pieces of planked cedar wood joined together using mortise-and-tenon connections, often employed in boat construction. One of the largest ancient ships is 43.4 meters (142 feet) long and 5.9 meters (19 feet) wide. The Khufu Ship is a magnificent illustration of ancient history.

Lesson: Investing in a Varied Fleet for Effective Transportation

Transportation was essential in ancient Egypt. Desert trade routes and the Nile River connected this ancient civilisation. This lesson will examine how ancient Egyptians employed various transportation techniques and fleets to support their trade, expansion, and progress. Invest in a diverse fleet of ships suitable for different trade routes and cargo sizes, ensuring efficient and cost-effective transportation.

1. Diverse Fleet for Different Trade Routes

Trade has boosted global economic growth and culture. Commerce drives economic progress, but no nation controls all resources. Thus, countries must always buy and trade with each other. Ancient Egyptians did this regularly. Egypt traded with other nations despite its wealth. Ancient Egypt could trade with Lebanon, Canaan, and Nubia in the fifth century BCE due to its abundant natural resources. Egypt's excellent position in the world contributed to its flourishing economy. Egypt remains in northern Africa. Greece was Egypt's northwest neighbour, and Egypt was the region's commercial hub. Mesopotamia, Persia, and Yemen bordered Egypt to the northeast and south. The ancient Egyptian economy relied heavily on its neighbours.

The Tale of Captain Hatshepsut:

A renaissance that began the New Kingdom period began when Hatshepsut ruled peacefully for nearly twenty years in the 15th century B.C. You may not have heard of her for a reason. Twenty years after her nephew Thutmose III took the throne, most of Hatshepsut's inscriptions and iconography were defaced or destroyed, her name and title were removed, and monuments in her image were vandalised, ostensibly part of Thutmose's political effort to erase his predecessor to legitimise his son's ascension. This left Hatshepsut forgotten for over three millennia until 20th-century archaeologists discovered her reign. Captain Hatshepsut recognised the need for a diverse fleet to accommodate varying trade routes. Her decision to invest in small river boats and large seafaring vessels allowed him to tap into multiple markets effectively.

2. Cargo Flexibility:

Ropes and nets restricted cargo on ships and carriages in antiquity. When conveying cargo, the ancient Egyptians used ropes of various diameters and nets. In addition to ropes and nets, the Romans employed leather straps to secure cargo on wagons and chariots. Railway progress in the 1800s led to freight restraint improvements. Steel chains and tarps to protect items from the elements became more common. After the vehicle was invented in the early 1900s, chains, ropes, nets, and tops were still used. Steel strapping was introduced in the early 1900s to supplement these restraint systems. Hay, cotton, and other agricultural goods were bound with steel strapping, invented in the late 1800s. Other industrial shipping applications grew to use steel strapping in the 1900s. Strength and durability made steel strapping suitable for restraining huge weights, especially long-distance loads. Due to its strength, durability, and low production cost, steel strapping became popular during World War II. Polyester strapping has replaced steel strapping in recent decades due to its safety, cost-effectiveness, flexibility, and ease of handling while remaining strong and durable.

The Papyrus Trade:

Egypt had papyrus, one of the few early hydraulic civilisations. Egyptians used the plant extensively, exporting paper and rope. Egyptian papyrus was the only paper supply for millennia. The author describes how paper sheets and scrolls were manufactured, codices evolved, production increased, and papyrus paper died. Ironically, Egypt no longer has this natural resource, and its souvenir papyrus comes from African species reintroduced. Egyptians used smaller boats to transport fragile goods like papyrus, ensuring their safe delivery. Larger ships, on the other hand, were reserved for bulkier commodities like grain and stone.

3. Trade Expansion:

Egypt was an extension of the spice routes that spanned Saudi Arabia, Oman, and India, and like Saudi Arabia, trade was crucial to the Egyptian economy under the Islamic caliphate. Egypt was an early glass producer and exported it through China via the Roads. Egypt and other Mediterranean nations taught Central Asians to make good in the 5th century. Egypt boasts an unmatched legacy, classic Nile cruises, oases, bazaars, commerce, and religious hubs.

The Expedition to the Land of Punt

Hatshepsut, a female pharaoh, undertook a mythical expedition to Punt in the 15th century B.C., recorded in a stone bas-relief. Track the Punt expedition using a realistic bas-relief line drawing in this interactive. Pharaoh Hatshepsut's expedition to the Land of Punt, using specialised ships designed for the journey, opened up new trade routes and established valuable trade partnerships.

Conclusion:

Ancient Egypt's varied fleet management is still a striking example of human ingenuity in the annals of history. Egypt could harness the force of its deserts and seas through the graceful Felucca and magnificent Khufu ship, which allowed for thriving trade and cross-cultural interchange. This varied fleet, which could adjust to the needs of various trading routes, represented the vitality of a society that depended on trade and communication. Egypt left a long legacy of prosperity and cross-cultural contact by expanding its influence far beyond its borders with cargo flexibility that matched the demands of a dynamic economy. The history of fleet management in ancient Egypt constantly inspires human development.

Chapter 3

SPECIALIZED TRADE

Ancient Egypt's contribution to the larger fabric of ancient civilisations is one of the few that stands out as particularly brilliant. Egypt, a land where the sands of the desert met the life-giving waters of the Nile, was a land that served as a crucible for the development of culture, commerce, and new ideas.

This chapter dives into the fascinating world of specialised trade in ancient Egypt. In this place, goods of immense significance came across borders and into the hearts of civilisations nearby. Trade was not only important to the ancient Egyptians because of its economic significance. Rather, it was a mechanism through which they could connect with the rest of the world and could share the treasures of their land. They were not merely merchants; rather, they were keepers of information and authorities in commercial affairs. This chapter will take you on a tour through Egypt's specialised trade, from local markets to worldwide hubs of exchange, using the renowned commodity of papyrus as a real-life example to illustrate the voyage.

Range of Commodities Traded Wood:

Ancient Egyptians had most of what they desired within their country's boundaries, but their environmental conditions prevented some goods from existing in a mass quality. Another instance of this is wood, which existed in poor quality and was the primary material utilised in creating ships, houses, and furniture. As a result, during the old kingdom, Egypt established a trading connection with the Byblos, Lebanon kingdom on the Lebanese coast, which ended up being one of its closest allies for two millennia. The ancient Egyptians brought cedar wood from other countries to use in constructing their naval ships, which were employed to defend the country from any outside dangers, such as sea people. They also imported various varieties of hardwoods from Africa, including ebony, a scented wood.

Metal:

Egypt did not have a plentiful supply of metal; the country only had a few gold resources and a limited amount of silver, iron, copper, and lead, which was insufficient to meet the country's demands for these metals. Egypt initiated the invasions of Egypt and the Sinia. It began exploiting its gold and copper mines, which led to beneficial developments and had ramifications on the international stage. Copper was transported to Ancient Egypt from Cyprus, and substantial amounts of gold were bartered with the rulers of Asia in exchange for their political support of the Egyptian kingdom. Cyprus was an important trading partner for the Egyptians. Asiatic copper was a bronze metal that was made naturally using tin. It was known as "Asian copper."

Jewellery and Precious Stones:

The ancient Egyptians had a significant fascination with precious stones as well as other forms of luxury items. Incense was made from the resins of myrrh, frankincense, and aromatic wood, all of which were mostly sourced from the country of the east. Certain items were also prepared using these ingredients. They obtained lapis lazuli from the Kush region in northern Bactria. Products like vegetable oils, eye paints, and cosmetics were imported from eastern Iran and Afghanistan. Alexandria was a major centre for producing glass goods throughout the Roman period. Many commodities, including turquoise, gold, agate, carnelian, and other precious jewels, were transported from the Persian Gulf to Egypt via the Oxus road. Alternatively, these materials were transported to Egypt via ships that sailed around the Arabian Peninsula or through the Canal of the Nile-Red Sea.

Cattle and Animals:

Ivory, ostrich feathers and eggs, leopard and lion skins, and other animal parts were among the many items that made their way to Egypt from western and southern Africa. Chickens were brought over from India in the newly established state, and camels became incredibly prevalent in Egypt during the Persian conquest. Horses appeared for the first time in the 13th dynasty, which was highly famous during the Hyksos. Other cattle and animals were brought into the ancient Egyptian kingdom at various times, such as when the Persians conquered Egypt. In addition, domesticated animals such as donkeys, goats, sheep, dogs, and cats were brought to Egypt at some point after Egypt's primordial foundations.

Agriculture and Crops:

Egyptians were the biggest exporter of grains in the distant past; yet, the country did not produce sufficient maise, so it had to be imported from the prehistoric nation of Lebanon. Dates were also exported and sold throughout the entirety of the Roman Empire. Long rolls of the papyrus were sold to customers. Egypt was the only country in the Mediterranean region that produced papyrus. Papyrus was a type of paper used as the primary writing medium in Europe until the Middle Ages. Egypt was also known to export artefacts, such as the sculptures and sarcophagi discovered at Byblos, located in the Levant region. Additionally, amulets, scarabs, and rings were discovered in Malta, along with beads from faience and torch holders. Egyptian artefacts such as weaponry, jewels, and mirrors were traded for frankincense, exotic woods, and ivory when the land of Punt was acquired.

Transportation:

During the Ptolemaic dynasty, the lighthouse of Alexandria was built to symbolise the tremendous worth and importance of shipbuilding and sea trade. The beacon was used to simplify and boost the goods transported from all of the corners of Africa; however, the ancient Egyptians also resorted to alternative methods of transporting their goods. The use of camels, caravanserais, and chariots, which made full use of the wheel technology developed between 1674 BC and 1549 BC, were the alternate techniques for traversing the eastern and western deserts, respectively.

Establishing Specialized Trade Centres

The stimulation of economic growth and the advancement of civilisation worldwide has been significantly aided by trade. Even while commerce is the engine that drives economic growth, no nation can claim monopoly status over all of the world's resources. Therefore, a country will always be required to make purchases and engage in trade with another nation. This was a standard practice in ancient Egypt. Even though Egypt possessed many resources, it still had to trade with other nations. Ancient Egypt was better positioned to trade with neighbouring countries such as Lebanon, Canaan, and Nubia in the fifth century BCE because natural resources were readily available. The prosperous state of Egypt's economy during that time was largely due to the country's advantageous position in the world. The location of Egypt has not changed; it is still in the northern part of the African continent. At this time, Greece was Egypt's immediate neighbour to the northwest, and Egypt was the region's focal point of commercial activity. Egypt was bordered to the northeast by Mesopotamia and Persia, and Yemen was to the south. Yemen was Egypt's southern neighbour. The ancient Egyptian economy owes much of its prosperity to the countries bordering Egypt then.

Ancient Trade Routes:

One economic and social event that Ancient Egypt participated in was trading. Transport was another important task carried out by Egyptians to ensure that products arrived at the locations for which they were intended. Because of this, the Nile River served as the hub that brought the nation's commercial sector to a cohesive whole. Along the journey of the Nile, ships and boats delivered various items to several trading ports along the river. After being unloaded, the items would be moved along every path by camels, carts, or people on foot. The Egyptians were eager to encourage their trade and ensure easy contact between the Egyptians and their Western and Eastern trading partners. Because of this, they constructed a canal that facilitated the passage of ships through the area more easily. Because of its close relationship with the early Egyptians, the kingdom of Kush was an important link in the chain that connected Egypt to most of Africa. The ancient Egyptians, Kushians, and Nubians all engaged in trade with one another, which likely explains why they had similar types of government, religious beliefs, and cultural practices. Pyramids were constructed by the Nubians, just like the Egyptians. In addition to this, they practised mummification and revered the Egyptian gods. Gold and iron were the most often exchanged commodities between the ancient Egyptians and the Nubians of ancient Egypt. Because of these parallels, communication between Egyptians and people from other parts of Africa was simplified.

From Local to International Trade:

The evolution of ancient Egypt's commerce networks has been extensively studied. Ancient Egyptian trade focused on agricultural surpluses. Annual Nile River flooding made the soil fertile for large harvests. These surpluses were traded in Egypt, boosting local markets and economic growth. Upper and lower Egypt had separate resources and specialities in ancient

Egypt. Trade between these regions included minerals, foodstuffs, and handcrafted goods. Regional trade boosted Egypt's economy and unity. Egypt's location aided international trade. The Mediterranean and Red Seas afforded natural trade routes. Egypt connected to the Indian Ocean via Red Sea ports like Berenice. Alexander the Great made Alexandria a Mediterranean commerce hub. This multicultural city facilitates trade between Europe, Asia, and Africa. Egypt was rich in natural resources and agricultural surpluses. These included gold, copper, jewels, and limestone, traded locally and globally. Egyptian artists made exquisite jewellery, linen, pottery, and cosmetics. International markets demanded these high-quality items. Papyrus writing changed communication and record-keeping. Papyrus scrolls were utilised for administrative, intellectual, and trade objectives. Advanced irrigation systems and a large canal network facilitated trade-related agricultural output and transit. Egypt's religious beliefs and magnificent buildings drew pilgrims from nearby places. Intercultural exchange often leads to trade relationships.

Specialised markets for papyrus

The papyrus (Cyperus papyrus), a sedge, was an important part of the ancient Nilotic environment and symbolically important to the ancient Egyptians. Dense papyrus thickets grew in Nile Delta marshes and low-lying places around the valley because they needed shallow freshwater or water-saturated ground. The slender, robust stalks with feathery umbels and little brown fruit-bearing flowers can grow five meters from a horizontal base. Papyrus might be employed to make a wide range of items. Skiffs formed by connecting the long stalks were employed for local transit and hunting beginning in the Predynastic period. As depicted in tomb art, the boats utilised during pilgrimages and funerals had these peculiar shapes; such may have been manufactured originally of reeds and were later translated into wood. The tough outer rind was removed to reveal a spongy white pith strengthened by long vein bundles that could be fashioned into durable strips for manufacturing, such as mats, boxes, baskets with flowers, lids, sandals, and ropes. According to Herodotus, the lowest section of the plant (presumably the root) could be roasted and consumed. Papyrus production is a lengthy and complicated procedure requiring competence in all stages, from plant growth and harvesting to roll production. Many researchers assume that this was a state-run activity, at least in later eras of Egyptian chronology; it has been proposed that the Greek term papuros originates from the Egyptian pa-per-aa, "that of the pharaoh," though there is no substantive evidence to support this. In any case, most likely due to a desire to reduce reliance on Egyptian supplies, parchment gradually surpassed papyrus as the most favoured writing material. The most recent known papyri were dated back to around 1100 A.D. However, these are isolated occurrences.

Conclusion

In conclusion, ancient Egyptian specialised trade left a lasting impact. It teaches us that trading involves sharing culture, craftsmanship, ideas, and things. It shows the transformative potential of invention and the persistent influence of commodities on nations. Papyrus and Egypt's specialised trade riches reveal a culture that enriched itself and left an everlasting mark on the world. Today, exchanging goods and ideas drives global trade and cultural development, echoing ancient Egypt. We leave antiquity with the teachings of specialised trade, a witness to human innovation and exchange.

Chapter 4

BUILDING STRATEGIC ALLIANCES

When the need for trade arose to export Egypt's surplus agricultural goods and import products that Egyptians were not self-sufficient in, the importance of trade relationships became evident.

But why trade relationships? Why not only trade? Trade relationships go beyond simple transactions; they're built on trust and the quality of goods exchanged. While trade is about exchanging goods, a trade relationship goes a step further. It involves building trust, reliability, and often long-term partnerships.

The Egyptians understood this well and nurtured strong relationships with neighbouring civilisations like the Levant, possibly the Phoenicians, Nubia, and Mesopotamia. They know the trading relationship is about creating a win-win situation for both parties. But it didn't stop there; ancient Egyptian traders also engaged in business with Anatolian and Punt, demonstrating their commitment to fostering trust and prosperous partnerships.

So, in this chapter, we'll delve into the Egyptians' profound understanding that by fostering trade relationships, they not only secured immediate gains but also laid the foundation for long-term prosperity and mutual benefits. We'll explore how they forged these invaluable alliances and with whom.

Trade Relationships with Neighbouring Civilizations (The Levant, Nubia, Mesopotamia)

In this section, we will analyse the trading relationship of ancient Egypt with its neighbouring countries, such as the Levant, Nubia and Mesopotamia. Let's start with Levant. Keep in mind that the Levant, encompassing modern-day countries such as Palestine, Lebanon, Jordan, and Israel, forms a significant part of our historical narrative. However, intriguingly, according to this study, ancient Egypt cultivated robust relationships primarily with the southern Levant. It's a testament to the enduring bonds that transcended time and geography. It means Israel, Jordan, Lebanon and the southern part of Syria. According to the Polish archaeologist findings, ancient Egypt boasted an impressive trade relationship with the southern Levant, a connection spanning over five millennia, beginning in the distant past. Initially sporadic, this trade link had its roots in the exchange of goods among nomadic tribes during their annual migrations.

Furthermore, Egypt's valuable resources found their way to significant locations on the Chalcolithic East Bank, like Teleilat Ghassul near the Dead Sea, and extended into the northern regions of Jordan. Remarkably, the southern Levant was an essential intermediary, facilitating these vital trade exchanges. Furthermore, it's fascinating to note that modern-day Lebanon was a source for importing cedarwood, a material believed to have been used to construct many iconic Egyptian monuments.

Not only this but as time advanced, it's evident that this trade relationship evolved into a more structured arrangement, possibly influenced by the rise of a prominent copper processing center in the Gulf of Aqaba. Some scholars even propose that the inhabitants of sites like Tell Hujayrat Al Ghuzlan, near Aqaba, were involved in producing copper ingots and established connections with Egypt's Nile Delta region. These interactions underscore the southern Levant's vital role as a pivotal partner within Egypt's ancient trade network. It emphasises how this region was crucial in facilitating the exchange of goods and resources between these two ancient civilisations.

Having explored ancient Egypt's relationship with the Levant, let's focus on its interactions with Nubia. Being close to the Nile River, Nubia naturally formed a significant part of Egypt's historical ties. Like many neighbouring connections, this relationship was a complex mix of cooperation and conflict. Both civilisations experienced periods of peace and war, with the Egyptians conquering Nubian lands on multiple occasions. It's a dynamic that reflects the ebb and flow of history between these two ancient powers.

The roots of their relationship stretch back to around 2000 BCE. During this time, Nubia became a vital source for Egypt, supplying gold, ivory, copper, tools, pottery, exotic spices, and stone vessels, especially around 2500 BCE. In return, Egypt reciprocated by providing Nubia with valuable grains and textiles. This mutually beneficial partnership enriched both lands, fostering a vibrant trade route along the Nile that contributed to the prosperity of both ancient. Moreover, according to the Egyptian transcript, Nubia featured a trade centre known as Yam; the precise location remains a mystery. However, as time passed, another trade centre named Irem took its place, particularly during the New Kingdom era. These transitions in trade centres signify the evolving dynamics of commerce in the region and the adaptability of both Nubian and Egyptian trade networks.

Now, let's journey into the relationship between Egypt and Mesopotamia. Trade with Mesopotamia is as ancient as civilisation, with its origins tracing back to the time of the first dynasty kings. These kings forged trading alliances with neighbouring nations and established a robust central government in Memphis. Similarly, Mesopotamia was among the earliest business partners of ancient Egypt. Their trading relationship was so significant that both cultures deeply influenced one another, leaving lasting marks on each other's histories.

In the realm of trade, during the Naqada II period, there existed a thriving exchange between both civilisations. At its core, this trade involved the transfer of arts, textiles, silver, and precious stones. Yet, it went beyond mere goods. Egypt also received a sea of knowledge, artistic styles, and technology from Mesopotamia. As a result, the trading alliance between these neighbouring civilisations wasn't just economically beneficial; it enriched Egypt's culture with a foreign influence and advanced technology from Mesopotamia, leaving a traceable mark on their shared history.

Economic and Cultural Exchange: The Profound Impact of Trading Relationships

There's no denying that when two nations exchange goods, it's far more than a simple transaction; it's a bridge for exchanging culture and knowledge. This phenomenon holds even greater significance in ancient times when people embarked on long journeys, allowing them to immerse themselves in their surroundings. For Egypt, this meant forging connections with brilliant civilisations that bestowed precious products and ushered in an era of cultural and economic prosperity. It's a testament to the profound influence that trading relationships can have on a nation's development.

A shining example of this phenomenon is the enduring bond between Egyptian and Nubian traders. Undoubtedly, their history is marked by forced treaties and conflicts, but it's equally defined by the deep connections they forged. This intricate relationship between these two civilisations is a testament to the complex tapestry of human interactions, where commerce and culture often intertwine unexpectedly.

Both civilisations openly travelled from one region to another, a testament to their interconnectedness. Additionally, Nubia faced a dry climate, leading many citizens to establish colonies in ancient Egypt. Beyond the exchange of goods, the relationship between ancient Egypt and Nubia is rooted in shared cultural values, highlighting the profound similarities that united these two great civilisations. For example, the early Egyptian pharaohs and Nubian kings shared remarkably similar symbols, reflecting their cultural ties. Furthermore, there's evidence of marriages between these two civilisations, suggesting that some pharaohs had Nubian origins. However, despite these cultural similarities, not all their trade interactions were peaceful; some were conducted through military campaigns, underscoring their complex and multifaceted relationship.

Hence, their trade relationship bore fruit in the form of remarkable Egyptian monuments, such as the illustrious Temple of Karnak and the grand Mortuary Temple of Hatshepsut. This trade enriched Egypt culturally and strengthened its economy, leaving a lasting legacy that continues to awe and inspire us today.

Likewise, this chapter's alliance between ancient Egypt and Mesopotamia takes centre stage. Their relationship transcends mere trade; it represents a profound exchange of knowledge and cultural influence. Many significant artefacts from both civilisations share striking similarities in design and structure, spanning architecture, art pieces, pottery, and even weaponry.

Moreover, their agricultural and animal husbandry practices bear the mark of mutual cultural influence. Furthermore, a strong resemblance in writing and alphabet usage is evident in both cultures, likely stemming from their frequent trades and trading colonies.

This influence is vividly illustrated in the early part of Egypt, highlighting the far-reaching impact of their interconnected histories. For instance, Egyptologist Margaret Bunson points out that the renowned Narmer Palette, dating back to the First Dynasty, shows distinct Mesopotamian design elements, evident in its depiction of creatures and intertwined long-necked serpents.

Lesson: Expanding Market Reach through Strategic Alliances

History bears witness to the wisdom of the Egyptian civilisation. Beyond their grand pyramids and monumental achievements, they displayed remarkable astuteness in forging alliances with other nations. Shamelessly, they leveraged these strategic partnerships to further the prosperity of their nation, showcasing not only their ingenuity in construction but also their prowess in diplomacy.

Now, let's delve into how the Egyptians leveraged their strategic trading relationships with neighbouring civilisations and the lessons we can extract from their practices.

- **Levant: A Strategic Alliance for the Flow of Goods**

With the Levant, the strategic alliance revolved around the flow of goods, particularly the importation of cedar and timber wood from Lebanon. This wood was crucial for Egypt's construction projects, as the nation primarily relied on agriculture and lacked significant wood production. The alliance with the Levant ensured a steady supply of this vital resource for Egypt's construction industry.

Given the abundance of construction projects in ancient Egypt that demanded wood, they recognised they could ensure a steady and plentiful supply of this vital resource by fostering a peaceful alliance with the Levant. This strategic partnership served Egypt's construction industry and solidified the bond between the two nations.

Hence, budding entrepreneurs can glean a valuable lesson from ancient Egypt. Building alliances isn't just about business transactions; it's about fostering relationships. Just as the Egyptians exchanged greetings on holidays, offered occasional gifts, and inquired about each other's well-being, modern entrepreneurs can follow suit to establish strong bonds with their business partners. This approach not only cultivates lasting friendships but also ensures the smooth and sustained flow of goods and business in the long term.

- **Nubia: A Golden Opportunity**

There was controversy surrounding Egypt's alliance with Nubia, but it became clear that Egypt greatly relied on imports from Nubia to meet its needs.

Beyond military campaigns, another strategic approach existed to forming alliances, as exemplified in Egypt's relationship with Nubia. Through this alliance, both regions exchanged knowledge about agriculture and effectively utilised the fertile Nile lands. This collaboration empowered Egypt in terms of agriculture and enabled them to become exporters of agricultural goods eventually.

The lesson here lies in the exchange of valuable expertise. In modern business, it's not just about trading goods but also about observing what makes other businesses thrive and succeed in their respective fields. Through exchanging knowledge, both parties can learn from each other and thrive together. It's a mutually beneficial approach to business growth and development.

- **Mesopotamia: A Mutual Learning Experience**

The alliance with Mesopotamia opened the doors to cultural enrichment for ancient Egypt. Mesopotamian technology brought innovation to Egypt and played a crucial role in constructing awe-inspiring monuments that continue to captivate the world.

It is said that around 3000 BCE, Egypt's First and Second Dynasties began incorporating recessed niches from Mesopotamian temples into their tombs. It remains uncertain whether Mesopotamian builders influenced Egypt directly or if Egyptian architects found inspiration from imported Mesopotamian seals. Additionally, the design of Mesopotamian ziggurats, originating around 5000 years ago, likely influenced the development of Egyptian pyramids, with the oldest pyramids appearing more than two thousand years later, strongly suggesting Mesopotamian cultural and technological influence on Egypt.

In many ways, the alliance between Mesopotamia and Egypt teaches us the lessons of tolerance and adaptation. They adopted each other's techniques and cultural values, leading to Egypt's prosperity and the creation of one of the Seven Wonders of the World. Modern businesses can also learn from ancient Egypt's example by embracing tolerance and learning from different cultures instead of rejecting them. This kind of learning can enrich their values and lead to greater success.

Conclusion

As concluded, if done correctly, ancient Egyptians had proven that trade could be a two-way road beneficial for economic and cultural growth. The Egyptians understood the significance of building trust and long-term partnerships, not just engaging in transactions. Through these alliances, they secured valuable resources like cedarwood from the Levant, exchanged agricultural knowledge with Nubia, and absorbed technological and cultural influences from Mesopotamia. Modern businesses can draw valuable lessons from Egypt's diplomatic and strategic approach to alliances, emphasising the importance of fostering strong relationships for mutual benefit and long-term success.

Chapter 5

EMPHASIZING QUALITY AND CRAFTSMANSHIP

From the elegant chalice to the everyday use of a razor blade, the artefacts unearthed by archaeologists provide compelling evidence of the exceptional quality that defined ancient Egyptian craftsmanship. It's truly remarkable to consider that these items have endured for thousands of years, a testament to both the skill of their makers and the superior raw materials used. Beyond this, the undeniable trading success of ancient Egypt stands as a testament to the enduring appeal and high quality of their goods.

Moreover, it's worth noting that the products they imported, such as wood from the Levant and gold from Nubia, were also of excellent quality. These materials were crucial in crafting their high-quality goods. In addition, the grains they cultivated, the discovered jewellery, and the linen used in mummification all testify to ancient Egypt's commitment to providing and using top-quality products rather than compromising with lower-quality alternatives.

However, there is a slight controversy surrounding this notion. It's possible that the artefacts found belonged to the elite and royal classes of ancient Egypt because many were discovered in pyramids and tombs of various pharaohs. This raises the possibility that common people in ancient Egypt might have used and produced lower-quality items that have not withstood the test of time.

Yet, it's important to acknowledge that the true merchants of ancient Egypt were its rulers. They held the responsibility for both domestic and international trade, ensuring the provision of high-quality products. This chapter will illuminate the exceptional quality of the products crafted by ancient Egyptian merchants subsequently traded with other nations or fellow citizens. We'll also delve into the valuable lessons that modern businesses can glean from their practices.

Highly Regarded Egyptian Goods

The great civilisations like Romans and Mesopotamia shared cultural similarities with ancient Egypt and a desire for Egyptian products. The presence of these Egyptian goods in these civilisations attests to their high demand. Romans and Mesopotamians were discerning consumers who sought only the best for themselves. This, in turn, fuelled trade between Egypt and these nations, contributing significantly to Egypt's economic prosperity.

Furthermore, consider the valuable items that were exchanged in these trades. Beyond textiles and grains, Egypt received precious commodities like gold and silver, which were relatively scarce in ancient Egypt. This further underscores the exceptional quality of Egyptian products, as people from other nations were willing to trade items as valuable as gold for them.

Now, let's embark on a journey to explore how ancient Egypt's remarkable craftsmanship, technological advancements, and quality materials have left behind awe-inspiring artefacts for archaeologists to discover.

Agricultural Products:

Thanks to the clever thinking of the ancient Egyptians, they found a way to make the most of the highly fertile land along the Nile River. They could even predict the river's annual floods, which allowed them to plan their crop planting effectively. Additionally, they pioneered a method known as basin irrigation, which helped them efficiently water their crops.

Most of the common people in Egypt relied heavily on vegetables as their primary food source. This was because meat was a costly and scarce luxury, typically enjoyed by the wealthy and elite classes regularly. The large-scale farming

practices of ancient Egypt played a crucial role in their economic prosperity. They cultivated a wide range of crops, including staple foods like wheat and barley and industrial crops like papyrus and flax. Additionally, they had gardens and orchards where they grew various fruits and even medicinal plants, including henna. This diverse agricultural activity contributed significantly to their economic well-being.

A fascinating aspect of their agricultural prowess is that even their food crops, such as wheat, became a significant export. Egypt produced an abundance of wheat, far more than they required domestically. The Romans were the primary importers of Egypt's wheat, and the supply was so plentiful that Egypt earned the title of 'Rome's breadbasket'. No matter whether it was distributing goods within their own country or exporting them internationally, ancient Egypt consistently upheld a standard of providing high-quality products. This commitment to quality was key to facilitating successful domestically and abroad trade.

Indeed, papyrus was a significant source of economic export for ancient Egypt. This plant was renowned for its use in making paper. Just imagine the paper from that era – intricate and delicate, yet of such quality that it ensured the longevity of the written word.

Byblos in the Levant played a major role as the primary exporter of this product, which was later transformed into paper. This paper was then exported to Mesopotamia and neighbouring regions. What's particularly fascinating is that the connection between Byblos and book production gave rise to the English word 'Bible.'

This historical interplay between hardworking and skilful agricultural techniques and producing high-quality goods is awe-inspiring.

Textiles:

Egypt is considered a pioneering civilisation in textile production, indicating that their weaving methods were likely distinctive to Egypt and had a notable impact on neighbouring civilisations such as Mesopotamia. Archaeologists have discovered luxurious textiles used in ancient Egypt, and it's remarkable to note that their textile-making process was done by hand.

They utilised flax plants to create linen, employing various techniques to transform the flax into yarn, skilfully woven into complete fabrics. In this intricate process, they relied on tools like netting needles and spindles to craft these remarkable textiles.

In addition to flax plants, the ancient Egyptians also utilised animal and palm fibres to produce wool and ropes. However, it's worth noting that Egyptians regarded wool as impure. Therefore, linen was esteemed as the preferred choice of clothing material.

The linen woven in ancient Egypt was characterised by intricate designs, exceptional craftsmanship, and unparalleled quality. This high-quality linen from Egypt was highly sought after, both within the kingdom and in trade with nearby civilisations. It's noteworthy that the civilisation of Mesopotamia was influenced by the weaving techniques of ancient Egypt.

The skilful craftsmanship of Egyptian textiles allowed them to command premium prices in the market, underscoring their reputation for excellence.

Indeed, one of the most compelling examples of high-quality linen from ancient Egypt is the linen used in the mummification process. While they employed spices and chemicals for preservation, the exceptional quality of the linen played a crucial role in its survival over thousands of years. Today, archaeologists can study these well-preserved linen wrappings, providing valuable insights into ancient Egyptian culture and craftsmanship. For example, the tomb known as Bab El-Gasus at Deir El-Bahari holds significant importance for studying ancient linen textiles. Inside this tomb, dating back to the Twenty-First Dynasty (around 1070–945 BC), researchers made remarkable discoveries of various linen burial fabrics. These pieces are crafted from high-quality linen, and what's truly remarkable is that their original colours remain intact to this day.

Jewellery:

While the ancient Egyptians did import metals due to a lack of high-quality local resources, they certainly did not lack a keen sense of intricate design and artistry when it came to their jewellery. The jewellery trade in ancient Egypt has a rich history dating back to the Predynastic period, which predates 3100 BCE. This demonstrates their enduring fascination with adornment and craftsmanship. During that era, ancient Egyptians utilised imported blue minerals from Afghanistan, known as lapis lazuli, in their jewellery. Their jewellery showcased exquisite craftsmanship and held cultural and social significance. These jewellery items were traded both within Egypt and with neighbouring civilisations.

One of the most outstanding examples of intricate craftsmanship in ancient Egyptian jewellery is the golden funeral mask of

King Tutankhamun. The burial mask of the young pharaoh serves as a remarkable showcase of the pinnacle of craftsmanship in the art of jewellery making. It is a testament to the extraordinary skill and artistry of ancient Egyptian artisans, revealing their ability to create unparalleled beauty and significance in jewellery.

Legacy of Ancient Egypt's Traders: Prioritizing Product Quality for Reputation

Indeed, there are a variety of valuable lessons that modern businesses can glean from ancient Egypt. From their commitment to high-quality craftsmanship to establishing a sterling reputation that would endure through the ages, ancient Egyptians set a remarkable standard for the world, including future generations, to aspire to.

Whether it was the creation of daily utility items or the crafting of intricate jewellery for pharaohs, ancient Egypt demonstrated that when you invest effort, skill, and quality raw materials into your products, you can achieve exceptional quality and build a successful business that brings prosperity to your land. Now, let's analyse the lessons modern businesses can draw from the ancient Egyptian quality standard for goods.

Craftsmanship and Quality of Material Matters

In business, success entails more than just financial gains; it also encompasses goodwill. This goodwill not only contributes to profitability but also shapes the character of a business and sets the standard. It plays a vital role in upholding the reputation of a business, a lesson we can discern from the enduring legacy of ancient Egypt.

Ancient Egypt's keen observation of craftsmanship and sharp understanding led to a crucial realisation. To build a sterling reputation for their businesses, they must maintain a steadfast commitment to delivering high-quality goods. As a result, they never compromised on the quality of their products. They invested in honing their craftsmanship skills, dedicated time and effort to perfect their creations, and utilised valuable resources to craft exceptional pieces for trade. This unwavering dedication to quality was a cornerstone of their success.

Can you imagine how it's truly remarkable to consider whether modern products could endure the test of time as effectively as the artefacts of ancient Egypt? The lesson here for modern businesses is clear: prioritise quality over quantity. By sharing their talent and dedicating themselves fully to their products before introducing them to the market, businesses can create enduring goods and build a legacy of excellence that stands the test of time.

Focusing on quality can help businesses build a strong market reputation. Additionally, offering workshops and training programs for employees to enhance their craftsmanship within the industry can be a strategic move. This investment in skill development not only ensures the longevity of the business but also solidifies its reputation in the market, making it a trusted source of high-quality products or services.

Meeting Demand

There is a variety of evidence in history indicating that the Egyptians were open to adopting innovative techniques and technology from Mesopotamian society. This flexibility and openness to new ideas illustrate their adaptability and willingness to embrace advancements.

Furthermore, their ability to research the market and cater to consumer demand is another noteworthy aspect of their approach. Over time, they adjusted their products to align with changing market demands and innovated accordingly. This adaptability and responsiveness to the market were crucial factors in their continued success.

While modern businesses have an array of tools to monitor consumer needs and market demand, there's always room for improvement. Therefore, they can take a valuable lesson from the history of Egyptian traders regarding the secret to a successful business: the importance of flexibility and responsiveness to changing demands. This historical wisdom serves as a timeless reminder of the critical role adaptability plays in achieving and sustaining success in the world of business.

Building a Reputation

As previously emphasised, high-quality products consistently earn a strong reputation for businesses in the market. It's inherent in consumer nature to gravitate towards businesses that offer such quality. When you're the one running a business, this principle holds true. Furthermore, high-quality goods often command higher prices, increasing business revenue. With a well-established reputation, you can confidently attach a premium price tag to your product, and it's likely to sell successfully. This symbiotic relationship between quality, reputation, and pricing is a key driver of business success.

In this regard, Apple Inc. provides a contemporary example of the principles observed in ancient Egypt. Regardless of the price, a segment of consumers is eager to purchase Apple's new products because the company has established a certain standard and elevated its reputation in the market. This lesson is valuable for modern businesses: by targeting specific markets and maintaining a strong reputation within those markets, they can earn long-term customer trust and expand their reach beyond national borders. This approach fosters brand loyalty and paves the way for international success.

Conclusion

As concluded, ancient Egypt placed a high value on the quality of their products and dedicated ample time to craftsmanship. This commitment earned them goodwill, customer trust, and increased profits. However, it's important to recognise that those were simpler times, and competition wasn't as intense as today. In the contemporary world, competition is fierce, but the fundamental values of business remain unchanged. These values include high-quality craftsmanship, dedication, skill, effort, and well-thought-out business strategies, all of which we observed in the traders of ancient Egypt. These timeless principles continue to underpin successful businesses today.

Chapter 6

STREAMLINING TRADE DOCUMENTATION IN ANCIENT EGYPT

Fake yourself again in time to the banks of the majestic Nile River, where the historical civilisation of Egypt thrived. In this bustling society, wherein lifestyles revolved around the river's fertile banks, trade wasn't simply a financial pastime; it turned into the lifeblood of society. The historical Egyptians possessed superb expertise in the importance of meticulous report-keeping in ensuring the easy flow of goods and offerings, and their early device of exchange documentation became a testament to their ingenuity and practicality.

The Heartbeat of Ancient Egypt: Trade

To sincerely respect the significance of the change in historical Egypt, one must first grasp the Nile River's primary role in their lives. This existence-giving waterway no longer only sustained the land with its annual floods but additionally served as a natural change course. The Nile linked regions extensively and extensively, facilitating the trade of products, thoughts, and traditions.

The bustling markets and harbours along the Nile were the epicentres of this exchange-driven civilisation. Whether it was grain from the fertile Nile Delta, precious metals from Nubia, or wonderful goods from the Red Sea, a huge array of commodities flowed through these markets. Trade became so intrinsic to Egyptian existence that it left an indelible mark on their subculture, artwork, and spiritual beliefs.

The Need for Meticulous Record-Keeping

With exchange permeating every aspect of lifestyles, the ancient Egyptians identified the imperative of preserving meticulous information. The key to fulfilling their trade operations lies in their capacity to appropriately record transactions, quantify goods, and make certain fairness in agreements. This desire for precision gave birth to a complicated machine for exchanging documentation.

Contracts: The Hieroglyphic Pacts

Contracts had been the bedrock of change in ancient Egypt. These agreements, often inscribed on papyrus, delineated the terms and situations of change transactions. Sellers and shoppers used this medium to file their expectations meticulously, specifying quantities, satisfactory requirements, delivery schedules, and even penalties for breaches. These contracts, adorned with hieroglyphs, were no longer simply felony devices; they had been artwork that tested the Egyptians' dedication to readability in their agreements. Ambiguity had no vicinity in these hieroglyphic pacts.

Imagine the scene as a scribe, seated, move-legged, meticulously etches out the terms of an alternate settlement on a roll of papyrus. Every stroke of the reed pen has profound significance. These files had been a sworn statement to the historical Egyptians' meticulous approach to trade documentation, ensuring that each event worried has been on the same web page, figuratively and actually.

Bills of Lading: Safeguarding Cargo on the Nile

As exchange flourished alongside the Nile, it became necessary to manipulate the difficult logistics of transporting items. To deal with this project, the historical Egyptians brought bills of lading, an essential component of their exchange documentation device. These files contained important information about the shipment, including its type, amount, and intended destination.

Imagine a bustling Nile port, with items being loaded onto boats destined for diverse places up and down the river. To ensure that the shipment reaches its vacation spot intact, payments of lading have been issued for each cargo. These historic payments of lading had been tangible clay drugs or papyrus scrolls, physical representations of alternate agreements. They served as a form of receipt, confirming the receipt of goods and their intended journey.

The Role of Hieroglyphs in Bills of Lading

Hieroglyphics, the ancient Egyptian writing machine, played a pivotal role in the bill of lading. Instead of relying completely on written language, the Egyptians used symbols and snapshots to provide statistics about the transported products. This visual machine-made trade documentation was available to a broader target audience and added a creative dimension to the process.

Each image on a bill of lading told a tale: the sort of items, the quantity, the vacation spot. It became a language understood by all, regardless of their literacy stage. This modern use of hieroglyphs in exchange documentation became emblematic of the Egyptians' creativity and their understanding of the strength of visual communication.

Inventories: Managing the Treasure Troves of Egypt

To keep control over their enormous shops and warehouses, Egyptians saved meticulous inventories. These statistics, painstakingly inscribed on scrolls or clay pills, catalogued every object in storage. From grains to textiles, treasured metals to pottery, nothing escaped the watchful eyes of the inventory keepers.

Facilitating Trade Negotiations

Inventories weren't mere lists of goods; they were strategic tools. When traders seek to strike offers, they may consult those inventories to ascertain the availability of specific commodities. Negotiations have been based on authentic facts, reducing the risk of overcommitment or underneath-shipping.

Responding to Demand

Additionally, inventories enabled Egypt to respond swiftly to changes in demand. If an unexpected surge in demand for a specific commodity arises, investors should consult the inventories to become aware of the available inventory. This agility in responding to marketplace dynamics became a testament to the Egyptians' ahead-wondering technique to change documentation.

The Role of Scribes in Inventory Management

The role of scribes in maintaining inventories cannot be overstated. These individuals possessed remarkable skill in record-retaining and would be tasked with meticulously updating inventory lists to indicate that goods had been obtained, disbursed, or eaten up. The scribes' accuracy became paramount because their work immediately encouraged trade selections, useful resource allocation, and universal economic balance.

Scribes, in many ways, had been the unsung heroes of historic Egyptian alternate documentation. Their determination to maintain correct inventories ensured that Egypt's big and complicated delivery chain functioned smoothly. It became a feat of organisational brilliance that laid the foundation for contemporary supply chain management practices.

In conclusion, as we step again into ancient Egypt's bustling markets and harbours, we witness a civilisation that understands the fundamental importance of trade and the essential position of meticulous trade documentation. Contracts, bills of lading, and inventories had been not simply bureaucratic methods; they were the threads that collectively weaved the fabric of Egyptian society.

These early change documentation practices were not necessarily for maintaining records. They were about permitting trade, safeguarding goods, and ensuring fairness in change. The historical Egyptians' commitment to accuracy and precision is a timeless lesson for modern-day companies, highlighting the necessary function of meticulous record-maintaining in facilitating trade and ensuring the smooth drift of products and services.

Lessons: Efficient Trade Documentation for Smooth Transactions

The ancient Egyptians' legacy in change documentation goes beyond ancient interest; it offers many lessons for current commerce. These instructions emphasise the significance of meticulous documentation to ensure clean transactions on a global scale.

Accuracy and Transparency: A Timeless Need

One of the essential classes from historical Egypt is the enduring significance of accuracy and transparency in change documentation. Just as hieroglyph-encumbered contracts left no room for ambiguity, cutting-edge organisations have to try for precise and transparent report-keeping. Ambiguities and misunderstandings can lead to disputes, delays, and economic losses.

Protection of Interests: A Principle That Stands the Test of Time

Ancient Egyptian exchange documentation became designed to guard the pastimes of all events involved. Contracts ensured that both shoppers and dealers were privy to their duties. In modern-day international trade, clear and complete contracts safeguard the rights and interests of businesses engaged in trade.

Efficiency and Accountability: A Common Goal

Efficiency and duty were additionally quintessential to the ancient Egyptian device. Bills of lading allowed for the green monitoring of products, reducing the threat of theft or loss. In modern-day global alternates, digital tracking structures and logistics software serve a similar motive, ensuring duty and well-timed transport.

Innovative Record-Keeping: Ancient Inspiration for Modern Solutions
The Egyptians' creative use of papyrus, clay capsules, and hieroglyphs to document statistics should encourage cutting-edge corporations to innovate their document-maintaining techniques. In the digital age, adopting modern-day generation for report management can notably streamline trade methods.

Efficient Trade Documentation in the Digital Age
The transition from papyrus scrolls to virtual files may additionally appear to have been a jump throughout millennia; however, the concepts of green exchange documentation remain timeless. In our modern-day world, the Egyptians' understanding can guide us in streamlining our change documentation methods.

Embracing Technology: The Digital Advantage
The cutting-edge generation gives us an array of technological gear to streamline alternate documentation. From digital contracts and virtual signatures to cloud-based stock control systems, the era can extensively enhance efficiency and accuracy. By adopting those tools, businesses can lessen paperwork, minimise mistakes, and speed up transactions.

Reducing Disputes: A Contemporary Challenge
One of the top goals of alternate documentation is to reduce disputes. Because the Egyptians used contracts to ensure clear agreements, contemporary companies can take advantage of nicely drafted contracts that leave no room for ambiguity. Disputes aren't the most expensive but can also harm relationships with trading companions.

Enhancing Traceability: A Key to Modern Trade
Traceability is a cornerstone of green alternatives. The Egyptians completed this through payments of lading, while modern-day organisations rely upon digital monitoring structures. These devices provide real-time visibility into the motion of products, enabling higher choice-making and faster responses to disruptions.

Minimising Delays: An Imperative in Today's Fast-Paced World
Delays in alternate can result in economic losses and strained relationships. By embracing green trade documentation practices, consisting of digital documentation and automated techniques, businesses can reduce delays due to guide office work, approval bottlenecks, or statistical discrepancies.

Environmental Considerations: A Modern-Day Concern
In modern international affairs, there's a developing attention to the environmental effects of change documentation. The historical Egyptians, using papyrus, also provided a lesson in sustainability. Crafted from the Cyperus papyrus plant, papyrus has become a useful renewable resource that might be harvested without depleting ecosystems. In comparison, present-day paper production regularly entails deforestation and a high energy intake.

Modern businesses can draw inspiration from the Egyptians by exploring sustainable options for standard paper-based documentation.

Cultural and Legal Aspects: Navigating the Complexities
The ancient Egyptians' approach to exchanging documentation was deeply intertwined with their subculture and criminal gadgets. Contracts, for example, had no longer been just legally

binding agreements; they often had non-secular importance as well. These contracts invoked the benefits of deities to ensure honest dealings. This mixing of spirituality with commerce highlights the profound connection between change and society in historic Egypt.

In the modern world, corporations engaged in international change must navigate a complex landscape of cultural norms and felony necessities. Understanding and respecting the cultural nuances of trading partners can lead to more fruitful relationships. Cultural sensitivity and cross-cultural communication are crucial in the modern, interconnected international marketplace.

Likewise, complying with worldwide changes in laws and regulations is essential to avoid legal headaches. Just as the Egyptians had their criminal framework for change, modern groups must operate within the bounds of global trade agreements, import/export policies, and contractual responsibilities. A breach of those prison parameters can result in disputes, fines, or sanctions.

Conclusion

The principles drawn from historical Egypt—accuracy, transparency, safety of interests, efficiency, and innovation in report-retaining—offer a solid foundation for present-day trade documentation practices. In an age in which worldwide trade is more complicated than ever, embracing those standards can result in smoother transactions, reduced disputes, more suitable traceability, and minimised delays.

Furthermore, as we look towards a more sustainable future, the ancient Egyptians' use of renewable materials, including papyrus, reminds us of the environmental impact of our documentation practices. Exploring eco-friendly alternatives to traditional paper-based documentation is not only accountable but also aligns with the values of modern-day society.

Chapter 7

MAINTAINING
STABLE CURRENCY

Maintaining a stable currency is vital to any country's economic strategy, impacting foreign trade and local markets. The stronger the purchasing power of a nation's currency, the more robust its economy becomes. Undoubtedly, the ancient Egyptians were an economically prosperous and affluent civilisation. Although there is no clear evidence of cash or a distinct monetary system until the 3rd millennium BC, it doesn't diminish the formidable purchasing power of money held by ancient Egyptian traders.

The evolution of currency in ancient Egypt is a fascinating journey, transitioning from grain to coins. Suppose you've ever wondered about the types of currency employed in ancient Egypt. In that case, this chapter will delve into the various forms of money they used, including a glimpse into the barter system.

Standardised Currency in Ancient Egypt

Regrettably, there is little evidence to suggest that ancient Egypt utilised any standardised form of currency apart from gold and silver rings. Over an extended period, Egyptian traders predominantly relied on the barter system for their local market transactions, primarily exchanging food items. When engaging with foreign traders, they dealt with more valuable commodities such as jewellery, textiles, and other precious goods.

In local trading, a term known as "mryt" referred to the riverbank, where market spaces were established along the Nile's banks. Here, people from across the country would gather to engage in barter trade, exchanging goods. Egyptians also conducted private trade amongst themselves. But there was a tradition in which the merchants bartering precious goods would publicly swear about the ownership and authenticity of their products. This system was effective in preventing fraud during that era. Now, let's delve deeper into the types of currency that ancient Egyptians used for their trade.

Barter System
Food was a significant demand in ancient Egypt and was often traded. However, the barter system was also utilised for various other items. The riverside market catered to local trade, where people engaged in small businesses dealing with daily necessities such as bread, beer, grain, textiles, and household items. Occasionally, animals were used in trading for more expensive goods. Additionally, historical evidence suggests that people even offered their daughters' hand in marriage as part of specific trade agreements. The most interesting fact is that, according to the Tehuti Research Foundation, merchants occasionally exchange a specific amount of metals to equate the value of the goods being exchanged. For example, if someone wanted to acquire a camel in exchange for a carpet, they might add metal nuggets to balance the value of the exchanged goods. As for wages, workers were typically compensated with food for their labour, which could vary depending on an individual's social standing. Therefore, the barter system remained prevalent throughout most of the era of the pharaohs in ancient Egypt, and it continued to be the primary system of trade well into the first millennium BC.

Grain, Bread or Beer as a Form of Currency
During the Old Kingdom, both the barter system and an emerging grain-based currency system coexisted. People began to seek alternatives to bartering, and that's when grain came into play. High-quality bread and beer were offered as wages to the labourers. Skilled workers, however, found it challenging to store large quantities of beer and bread, so they started requesting grain as an alternative form of payment. This gave rise to the concept of the grain bank, essentially a grain warehouse where ancient Egyptians stored their surplus grain. They could withdraw grain from these banks when needed since storing large quantities at home was impractical. This same method was applied to trade, making grain (and often oil) a de facto currency, complete with its banking system.

Gold, silver and Copper rings
The various forms of currency mentioned earlier were predominantly used for local trade within ancient Egypt. However, the barter system still played a significant role in foreign trade and interacting with merchants from other regions. Foreign trade required a means of exchange, where the value of these commodities and forms of payment would have been especially important.

Ancient Egypt's approach to currency was unique in that they relied on receiving money from other civilisations, like Greek merchants, rather than minting their coins. This was partly due to the scarcity of high-quality metals in Egypt. When they did receive metallic currency, they often didn't use it as coins but as ingots. For instance, the coins found in the treasure of Amarna were in the form of melted or powdered silver, primarily intended for industrial purposes rather than direct use as currency. This highlights the creative ways ancient Egyptians adapted to their circumstances for trade.

From the 14th century to the 4th Century BCE, a new development emerged in ancient Egypt's currency system: the use of metal rings. However, these metal rings weren't a form of money in the conventional sense. Instead, they served as weight units and were known as "dbn," vocalised as "deben," which translates to "rings." These deben rings played a crucial role in trade by standardising weights and measures.

Undoubtedly, while deben rings weren't traditional coins, they served to measure and establish the value of goods in ancient Egypt. The use of deben rings brought uniformity and predictability to trade. For instance, if one sheep was valued at three debens and one bottle of beer at one deben, merchants could easily barter these items, exchanging three bottles of beer for one sheep, thanks to the standardised value provided by the deben rings. This system helped streamline trade and ensure fair exchanges.

The value of Metal Rings or Debens

The "Debens" were crafted from various metals, from gold to copper. While copper and bronze held the same standard value, a distinct situation emerged with gold and silver.

Remember that silver was valued more than gold because it was a rarer metal during the Old Kingdom. However, when it comes to Debens, they were used for weighing, so gold weighed less than cheaper metals like copper or bronze. Let's revisit silver, known as "hedj", in ancient Egypt. Interestingly, this word later also came to refer to money. During the late Middle Kingdom, silver began to be used as a unit of weight, but it was used a bit differently from Debens. In ancient Egypt, they often crafted flat silver discs known as "senyu." These discs weren't used for their weight but rather to determine the value of items.

Other forms of currency

In addition to the previously mentioned forms of currency in ancient Egypt, evidence suggests using other types of money for trade. However, this evidence remains somewhat scarce and insufficiently documented. One example is the period preceding the emergence of local currencies when the flow of cash and ingots in ancient Egypt primarily represented the intrinsic value of the metal itself, particularly silver, in the context of barter. While indications of valuing minerals date back to the Old Kingdom, such evidence became more prevalent during the New Kingdom, notably in the records of labourers in Deir el-Medina. However, whether this practice was widespread in markets far from Deir el-Medina remains uncertain.

Professor Maspero observed scenes within a market where individuals exchanged small boxes containing mysterious contents. He postulated that these boxes might have held pieces of metal functioning as a form of currency, drawing this conclusion from his meticulous scrutiny of the depicted scenes. These boxes probably contained metal fashioned into small ornaments or alloys, adhering to standardised weights. This explanation aligns well with their appearance in market scenes, especially considering the absence of other items being exchanged by those carrying such compact boxes.

Another form of currency in ancient Egypt was known as "Sha'at," but, as in the case mentioned earlier, there isn't sufficient evidence to definitively classify it as a form of money.

In 1910, an inscription discovered by Stepptendorf in the Guizeh cemetery, dating back to the time of King Cheops, referenced this currency. It was used to sell a house from Teneti to Kambo, with Kambo paying ten Sha'at for the house, which included various items like wooden furniture and a cedar bed. The exact meaning of Sha'at remains a subject of debate, with some suggesting it was a unit of measurement or weight for precious metals used in pricing. However, it appears this currency had limited use in ancient Egyptian civilisation and was not widely adopted as a means of exchange. Interestingly, related derivatives of Sha'at, such as Shenat, Sheena, and Sinu, have sparked discussions about their commercial significance, although clear conclusions remain elusive. Therefore, while there may have been other forms of currency in ancient Egypt, more evidence is needed to provide reliable information on this topic.

From Barter to the Introduction of the 'Stater'

Over time, the true currency of ancient Egypt began to take shape as trade flourished, and Alexandria's port became a hub for foreign merchants and soldiers. During this time, the need for a standardised form of money became evident. Furthermore, as other Mediterranean civilisations like the Romans and Greeks adopted their currencies, there was a growing demand for money in exchange. It was during the 30th Dynasty, around 360 BCE, that Egypt introduced its first official gold coin, known as the "stater." Pharaoh Teos authorised the minting of this coin, primarily to compensate Greek mercenaries and other foreign workers employed by the state, marking a significant development in Egypt's monetary history.

Lesson: Ensuring Stable Currency for Seamless Transactions

Even during the Old Kingdom and using the barter system, ancient Egypt provided valuable lessons for modern traders. While we now rely on standardised currency, it's evident that they developed numerous traditions and methods to establish trust in their transactions and ensure peaceful trade.

Modern businesses can draw inspiration from ancient Egypt's emphasis on Standardization for Trade Clarity. They conducted clear and equitable trade without a formal currency, whether by offering high-quality grain as wages or using Debens. This commitment to fair practices was a key reason foreign traders had confidence in their currency, leading to smooth and trouble-free exchanges. Ancient Egypt's approach to trade offers timeless insights into building trust and transparency in commerce.

Conclusion

As concluded, ancient Egypt demonstrated a remarkable ability to adapt and evolve its currency systems as needed. Their creative methods for measuring the value of products, such as Debens or even a bottle of beer, are noteworthy. These insights shed light on the resourcefulness of Egyptian traders and their utilisation of diverse and intriguing forms of currency to facilitate fair and trustworthy trade relationships. Ancient Egypt's economic history offers valuable lessons in flexibility and ingenuity that continue to resonate today.

Chapter 8

INVESTING IN INFRASTRUCTURE:
THE BACKBONE OF ECONOMIC PROSPERITY

In the grand tapestry of human history, one of the greatest threads that weaves through time is infrastructure. It's the silent architect behind our societies, the cornerstone upon which civilisations have risen and thrived. Infrastructure isn't always pretty much bricks and mortar; it is approximately connections that have formed how we stay, work, and coexist. Imagine a time when nomadic hunter-gatherer tribes roamed the earth. They had been the developers of their generation, crafting tricky monuments like barrow mounds, stone tombs, geoglyphs, and stone circles. Yet, despite their skill, constructing roads wasn't a concern. Why? Because those nomads thrived on the open land, and their herds needed the liberty to roam.

But the whole thing was modified with the Agricultural Revolution. Settled groups sprouted from the fertile soil, and suddenly, infrastructure took on a brand-new significance. It has become the lifeblood of these rising societies, enabling them to exchange, journey, and collaborate. Early roads have not been grand creation initiatives; however, as a substitute, they result from centuries of human and animal footprints. Imagine well-trodden paths etched into the earth's canvas, winding their way between settlements. These paths were the precursors to urban planners now call "desire paths." Choice paths have been the organic infrastructure of their time, connecting groups and allowing the drift of goods, thoughts, and lifestyles. They have been the early manifestations of human ingenuity, a testament to our innate pressure to forge connections and overcome geographically demanding situations. As we journey again via the annals of records, we see that infrastructure has usually been the conduit of development. From humble desire paths to the complex networks of the current global, it's a tale of human collaboration, edition, and innovation.

Infrastructure isn't always pretty much roads and bridges; it is approximately the bonds that those structures create. It's approximately the communities they connect and the opportunities they release. It's the tale of how humanity, at some stage in the past, has tirelessly worked to bridge the gaps that separate us, turning nomadic trails into global highways.

 Imagine status on the bustling waterfront of an ancient harbour, the salty breeze to your hair, and the remote call of seagulls filling your ears. It's a scene that inspires images of adventure, trade, and exploration. But this romantic vision is more than simply nostalgia; it's a testament to the long-lasting importance of harbour infrastructure and trade routes in shaping the destinies of countries.

In this chapter, we embark on a journey to explore the charming world of infrastructure investment, with a special cognisance of the crucial roles performed using harbours and trade routes. We will also examine the significance of infrastructure for economic prosperity, harbour infrastructure and trade routes, supporting ships and caravans, and what lessons we will research from historical infrastructure in the present.

Harbour Infrastructure and Trade Routes

Picture a majestic sailboat, its towering masts reaching for the sky, as it procedures through a bustling harbour. Laden with exclusive spices, silks, and valuable metals from remote lands, it is an image of the vibrant global change that has shaped our international community for centuries. But these maritime adventures would not be possible without the problematic dance of harbour infrastructure and changing routes.

Harbours are the gateways to the world's oceans, where cultures meet, items are exchanged, and tales are woven into the tapestry of records. Trade routes, whether or not the historical Silk Road or modern-day highways, are the veins and arteries of monetary life, connecting disparate regions and fostering a sense of shared humanity.

In the world of trade, time is money, and efficiency is paramount. Efficient harbour infrastructure ensures ships can dock, sell-off, and reload with the precision of a nicely choreographed ballet. Likewise, well-maintained change routes assure secure and swift passage for caravans and vans. This efficiency isn't only a matter of convenience but a linchpin for the monetary boom.

The Rich Tapestry of History: Tales from Harbour Cities

To surely respect the importance of harbour infrastructure and change routes, we should embark on an adventure through time, exploring some of the sector's most iconic harbour cities and their contributions to global exchange.

Alexander, Egypt:

A Mediterranean Jewel In the historic world, Alexandria was a beacon of civilisation on the Mediterranean coast. Founded with the assistance of Alexander the Great, this city featured a thriving harbour at the crossroads of Europe, Asia, and Africa. However, its renown extended beyond just its geographic positioning.

Alexandria was additionally home to one of the Seven Wonders of the Ancient World—the Pharos lighthouse. This towering wonder guided sailors properly to its beaches and stood as a testament to human ingenuity. The town's bustling harbour has become a melting pot of cultures and commerce, where goods from across the globe have discovered their way into keen hands.

Lesson: Infrastructure Development for Efficient Trade
The training from records is clear: funding in infrastructure is a key motive of the monetary boom. Let's take a more in-depth examination of the training we will draw from the investment in harbour infrastructure and alternate routes, in particular from the angle of Egypt.

Investment in Infrastructure: Egypt's Success Story
With its strategic location at the crossroads of Africa and Asia, Egypt has a rich history of alternate relationships dating back to historical times. The pharaohs recognised the importance of green change and invested in harbour infrastructure and alternate routes alongside the Nile River.

The Suez Canal:
Among Egypt's most famous infrastructure endeavours is the Suez Canal. This synthetic waterway hyperlinks the Mediterranean Sea to the Red Sea, substantially shortening the voyage for ships journeying between Europe and Asia. The Suez Canal has served as a crucial direction for global change, and Egypt's investment in it has yielded fantastic returns.

The Nile River:
Egypt's prosperity has constantly been closely tied to the Nile River. The river served as a herbal alternate route, and the construction of canals and ports alongside its banks strengthened its significance. The Nile facilitated the motion of products from southern Africa to the Mediterranean coast, fostering alternate and monetary increases.

Trade Hubs:
Ancient Egyptian cities like Alexandria and Memphis have become thriving exchange hubs thanks to their strategic locations and properly advanced harbour infrastructure. These cities attracted traders from all corners of the globe, creating a melting pot of cultures and trade.

Lesson: Prioritise Investments in Infrastructure Development
Egypt's success in investing in harbour infrastructure and changing routes provides treasured training for current countries.

Connectivity is Key:
Just as Egypt leveraged its geographical gain with the Nile River, nations nowadays have to prioritise connectivity. Invest in transportation networks, along with roads, railways, and airports, to ensure the green motion of goods and people.

Embrace Innovation:
The Suez Canal became a groundbreaking engineering surprise of its time. Modern infrastructure initiatives must include innovation and technology to boost performance and decrease environmental effects.

Strategic Planning:
Egypt's funding for alternate routes and ports has become no longer haphazard. It became a part of a larger strategic plan to boost trade and financial growth. Nations have to increase their complete infrastructure development plans that align with their financial goals.

Sustainability Matters:
As we navigate the demands of the 21st century, sustainability should be a guiding principle in infrastructure improvement. Green infrastructure projects can benefit both the financial system and the environment.

Global Perspective:
Egypt's fulfilment in attracting merchants worldwide underscores the worldwide nature of exchange. Nations should view their infrastructure investments in an international context, thinking about how they can participate and enjoy the interconnected world economic system.

The Silk Road: An Ancient Network of Prosperity
To sincerely appreciate the importance of harbour infrastructure and change routes, we need to journey back in time to explore one of history's most iconic change networks—the Silk Road. This big community of interconnected routes spanned Asia, the Middle East, and Europe, facilitating the change of products, lifestyles, and thoughts.

The Birth of the Silk Road:
The Silk Road changed into not a single road but a complicated web of interconnected routes. It earned its name from the moneymaking alternate in silk, which was fantastically admired inside the West.

Beyond cloth wealth, the Silk Road facilitated the alternation of cultures and ideas. Philosophies, religions, art, and science travelled alongside those routes, enriching the societies they touched. Although the Silk Road's heyday is past, its legacy endures. Today, nations seek to revive and expand this historical change route through the ambitious Belt and Road Initiative, connecting Asia, Europe, and Africa via a contemporary infrastructure network.

Conclusion
As we stand on the edge of the 21st century, investing in infrastructure, particularly in ports, transportation, and trade routes, can't be overemphasised enough. Egypt's historic accomplishments, consisting of the Suez Canal and the Nile, are enduring proof of the undying fee of a sustainable transportation and exchange community.

Infrastructure, whether historic harbours or modern transportation hubs, is the inspiration upon which societies are built. It is the path to prosperity, the conduit of tradition, and the engine of innovation. In our ongoing journey of development, as we continue to invest in the infrastructure of the following day, we aren't simply constructing roads and bridges; we are constructing the very scaffolding of our shared human tale.

Conclusion

Ultimately, the legacy of shipping and trading practices of ancient Egypt isn't just a relic of the past but a beacon guiding us in the direction of performance, transparency, and sustainability within the cutting-edge world of commerce. How they conducted themselves in trade, their treatment of consumers, their ability to devise effective bargaining methods without formal coinage, and their innovative approaches to shipping and document handling in ancient times were truly remarkable. Moreover, by weaving collectively the know-how of the ancients with contemporary generation and high-quality practices, we can navigate the complex seas of worldwide trade with self-belief, ensuring that our transactions float as smoothly as the historic Nile, developing a more prosperous and greener worldwide marketplace. Therefore, these valuable insights have the power to steer modern businesses towards refining their shipping and trading approaches, nurturing expansion, and seizing the full potential of trade opportunities while dealing with difficult situations.

So, the next time you gaze out at a bustling harbour or traverse a well-maintained change route, remember that you aren't simply witnessing the motion of products; you're witnessing the heartbeat of humanity itself, where cultures meet, thoughts collide, and the world actions forward, one dock and one avenue at a time.

FINAL SUMMARY

For the better part of the modern age, the general stereotype around the Egyptian era is based on rich kings, with pyramids and "mummy-making" as the main occupations. Contrary to that Hollywood-influenced belief, the Egyptian Empire was like no other.

From 3100 B.C. to approximately 332 B.C.E., they made progress by relying on ingenuity and intellect instead of modern technology. Given their proximity to the Nile River and the Mediterranean Sea, they established an economy that greatly used the waterways to improve as a people. These improvements were comprised of but not limited to cargo ships, military boats and funeral vessels, as it was believed that one could only enter the afterlife via boat.

Trade and trade relationships were of significant importance because it was a mechanism through which they could connect with the rest of the world and share their land's treasures so much that they were considered Rome's breadbasket.

The Egyptians also introduced one of the first forms of paper derived from the papyrus plant. However, instead of relying completely on written language, the Egyptians used symbols and snapshots to provide statistics about the transported products.

Ultimately, the legacy of shipping and trading practices of ancient Egypt aren't just ancient relics. These insights can guide modern businesses in refining their shipping and trading approaches, nurturing expansion, and seizing the full potential of trade opportunities while dealing with difficult situations.

www.ingramcontent.com/pod-product-compliance
Lightning Source LLC
Chambersburg PA
CBHW080852120626
46546CB00008B/2798